Printed in Mexico

ISBN 978-0-15-362271-7
ISBN 0-15-362271-7

1 2 3 4 5 6 7 8 9 10 050 16 15 14 13 12 11 10 09 08 07

Visit *The Learning Site!*
www.harcourtschool.com

The Milky Way

Earth is our planet. But it is just one body within a large universe. Earth is part of a huge system of stars called a **galaxy**. Our particular galaxy is called the Milky Way. Not only are there billions of stars within the Milky Way, but there are also billions of other galaxies in space. Most of them are so far away from Earth, that they can't be seen without strong telescopes. There are many stars within our own Milky Way that aren't even visible without powerful telescopes.

Although we can't see it from Earth, the Milky Way galaxy is a spiral shape with a center and huge curved arms made up of stars and gases.

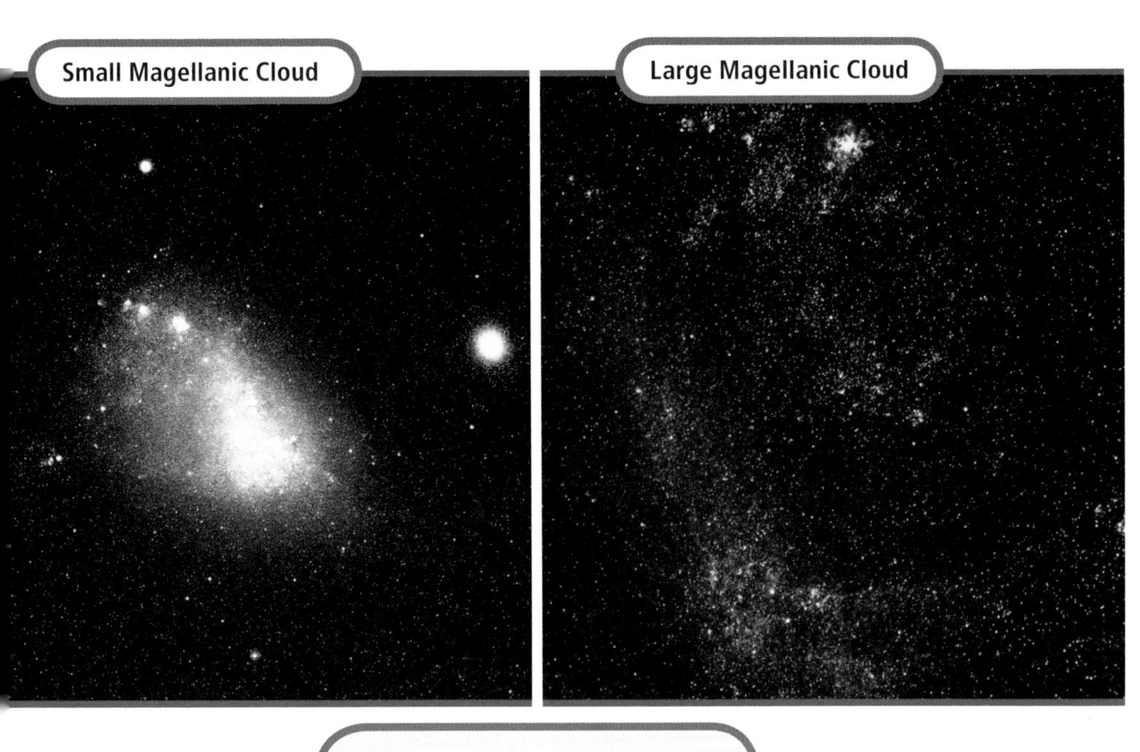

Even without the sophisticated equipment of today, Magellan was able to identify these galaxies.

Galaxies have different shapes. Some are elliptical, or irregular. Others, such as the Milky Way, have a spiral shape. About 20 percent of all galaxies have this shape. Most galaxies have an elliptical shape. They are small and flat, and some of them rotate.

A small percent of galaxies are irregular in shape. The stars that make up these galaxies have no set form or pattern. Two of these irregular galaxies are the closest neighbors to our Milky Way. They are called the Large Magellanic Cloud and the Small Magellanic Cloud. The early Portuguese explorer Ferdinand Magellan saw them on his voyage around the world in 1519.

MAIN IDEA AND DETAILS **Describe two different types of galaxies.**

Our Solar System

Billions of stars are only part of what makes up the Milky Way. Our solar system is also part of the giant Milky Way galaxy. Like our own solar system, there may be many planetary systems circling stars in the Milky Way.

The major parts of our solar system are the sun, nine planets, and their moons. All nine planets revolve around the sun. Some planets are so close to the sun that they are too hot to sustain life. Others are so far from the sun that they are freezing cold, and also unable to sustain any kind of life. So far, scientists know that Earth is the only planet that receives the right amount of heat and light from the sun in order to have life. It also has the right amount of oxygen and atmosphere to allow living things to live, breathe, and grow.

Astronauts aboard the International Space Station can remain in space for long periods of time. The space station is equipped with oxygen, water, and other life supporting systems like those on Earth.

Earth's moon. Pluto is the other planet in our solar system with only one moon. Saturn has many moons—at least 35!

Moons also fill the solar system. There are 141 known moons within our solar system. A moon is actually a satellite. A **satellite** is a body that revolves around another, larger body in space. The earth has only one moon. Its surface is rough and rocky. Mercury and Venus have no moons, whereas Jupiter has 62 that have been identified. Even Pluto, the smallest planet in the solar system, has a moon called Charon.

MAIN IDEA AND DETAILS **What are some of the characteristics of the earth that allow it to sustain life?**

Fast Fact

When astronauts planted the American flag on the moon, wires were put inside the flag to make it look as if it were waving in the breeze. There is no atmosphere on the moon, so there is no wind to make it wave.

The Planets

The planets in our solar system are divided into two groups. Mercury, Venus, Earth, and Mars are closest to the sun. These are called the inner planets. The inner planets are dense and rocky. Each one has layers consisting of a core, mantle, and crust. The inner planets also have atmospheres. However, Earth is the only one whose atmosphere can maintain life. Mercury's atmosphere is 100 times thinner than that on Earth, and Venus has an atmosphere 90 times thicker than that on Earth.

Jupiter, Saturn, Uranus, Neptune, and Pluto are called the outer planets. They are farther from the sun. The first four outer planets don't have rocky surfaces as the inner planets do. These outer planets are made up of gases and really have no solid surface at all. Pluto, though, has a rocky or icy surface. Another planet may exist even farther away than Pluto. Scientist have named this object Sedna.

Even though Venus is the sixth largest planet in our solar system, it is tiny compared to the sun.

How Long Is a Year?	
Planet	Year Compared to 1 Earth Year
Mercury	1 year = 88 Earth Days
Venus	1 year = 225 Earth Days
Earth	1 year = 365 Days
Mars	1 year = 687 Earth Days
Jupiter	1 year = 12 Earth Years
Saturn	1 year = 29.5 Earth Years
Uranus	1 year = 84 Earth Years
Neptune	1 year = 165 Earth Years
Pluto	1 year = 248 Earth Years

One year is the amount of time it takes a planet to revolve around the sun. Mercury is so close to the sun that its year is only 88 days. Pluto is so far from the sun that almost 250 Earth years make just I year on Pluto.

In 2006, scientists defined a planet as a sperical body that orbits the sun and clears its orbit. Pluto was reclassified as a "dwarf planet" because it does not have a cleared orbit.

Each planet has its own particular atmosphere and temperature. Jupiter is the largest planet in our solar system. It is more than 300 times bigger than Earth. Its atmosphere has powerful winds and lightning. Mercury, the second smallest planet in the solar system, has a surface similar to that of Earth's moon. Its temperature ranges from –183ºC (–361ºF) to 427ºC (801ºF).

CAUSE AND EFFECT **How does the distance from the sun affect the amount of time in a planetary year?**

Asteroids and Meteors

The solar system has planets that rotate and orbit around the sun. It also has chunks of rocks and metal orbiting the sun too. These are called asteroids. An **asteroid** is a chunk of rock and metal that orbits the sun but is too small to be a planet. Most asteroids are located in the area between Mars and Jupiter. There are so many asteroids in that area, that the band is called the asteroid belt.

Most asteroids are only about 1 kilometer (about 0.6 mile) in diameter. Ceres is the name of the largest known asteroid. Its diameter is about 1,000 kilometers (620 miles).

A meteorite formed this crater in Arizona 25,000 years ago. The hole is more than 1.3 kilometers (.75 mile) wide.

Sometimes chunks of rock enter Earth's atmosphere and burn up because of the friction. These are meteors. A **meteor** is a small chunk of rock, smaller than an asteroid. Most meteors form when asteroids collide. As a meteor falls through Earth's atmosphere the friction heats the rock as well as the surrounding air. This produces a trail of bright light. Even very small meteors can produce large, bright, trails of light as they fall through the atmosphere. Sometimes people refer to meteors as shooting stars, but they are not stars at all. Most meteors burn up before reaching Earth, but some actually survive and land on Earth's surface. These are called meteorites.

Fast Fact

Unlike the other outer planets, Pluto is not made of gases. It's mostly made of rock and ice. Some astronomers hypothesize that because of this, Pluto is actually not a planet but an asteroid.

SEQUENCE How does a meteor form?

Asteroid Discovery Dates

Asteroid	Diameter	Discovery
Aten	1 km	1976
Apollo	1.4 km	1932
Gaspra	16 km	1916
Juno	246 km	1804
Eunomia	272 km	1851

Comets

A **comet** is a large ball of rock, ice, and frozen gases that orbits the sun. One comet can have a very wide orbit. It can move close to the sun and move out far past Pluto. Each time a comet gets close to the sun some of its ice melts. The heat releases a cloud of dust, which forms one of a comet's tails. No matter how and where the comet moves, its tail always points away from the sun.

More and more ice melts off the comet each time it goes near the sun. Over time it breaks up and these smaller pieces move along the comet's former orbit. Some of these pieces eventually move into Earth's atmosphere as meteors.

The end of this large cloud of dust always points away from the sun.

Over the years people have given names to some comets. They look for these particular comets as they reappear in the sky many years after they were first seen. Halley's comet can be seen every 76 or so years. It was seen in the year 1066 as a spectacular glowing object in the sky. Over hundreds of years it has lost a lot of its rock and ice. The last time it was seen, in 1986, it was very dim and barely visible without a telescope.

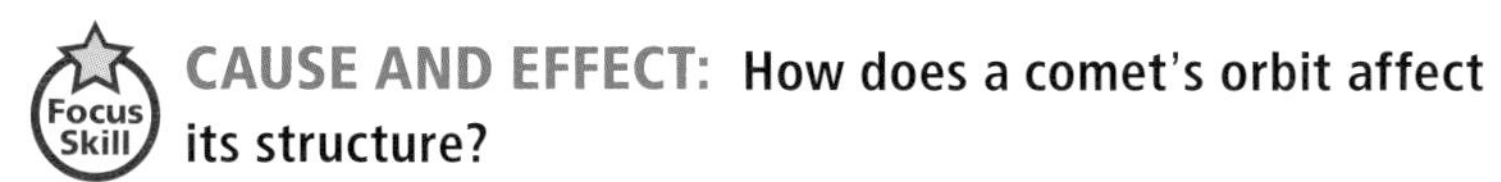

CAUSE AND EFFECT: How does a comet's orbit affect its structure?

Halley's comet is expected to be visible from Earth again in the year 2061.

The Sun and Stars

The sun is a star. It is closer to Earth than any other star in the solar system. Like all stars, it is made up of hot glowing gases such as hydrogen and helium. The energy of all stars comes from a process called fusion. This is the process in which the nuclei of atoms join and give off tremendous amounts of light and heat energy. Great amounts of pressure and heat are at the center, or the core, of a star. These are what make fusion possible. By studying the sun, scientists have been able to learn more about stars in general.

Astronomers classify stars by their color, brightness, and surface temperature. The stars with the most mass are hotter and brighter than those that are smaller. Cooler stars are red. The hotter stars are blue. Medium stars, such as our sun, are yellow.

In this picture, the cooler stars are red. The hotter stars are blue.

There are two different ways of measuring the brightness of a star. One way is by measuring the amount of light a star gives off, regardless of how it looks to us on Earth. This is called absolute magnitude. Some stars may look rather dim from Earth, but they give off an enormous amount of heat and light. They just look dimmer than some others because they are so far away from us. Another way scientists measure the light from a star is called apparent magnitude. This is how bright the star appears from Earth. Sirius appears as the brightest star in the sky other than the sun. Its absolute magnitude is actually 23 times brighter than the sun.

MAIN IDEA AND DETAILS What are the colors of the stars as they relate to their temperatures?

When Stars Die

Stars of all different ages shine within the Milky Way. Stars change as they grow older. How they change depends on their mass. Stars that are small use up their hydrogen slowly and can last more than 30 billion years. When they finally use up their hydrogen, they become white dwarfs.

Medium-sized stars last about 10 billion years. When they use up their hydrogen they become red giants. Huge stars have the shortest lives but they still live for millions of years. When they use up their hydrogen, they swell into supergiants and then explode. If any part of the core survives the explosion it collapses and forms a small, dense, neutron star. A really tremendous star can produce an even bigger explosion. This can then form a very dense black hole. The gravity within a black hole is so strong that nothing can escape it — not even light.

Focus Skill **SEQUENCE** **Describe the formation of a black hole.**

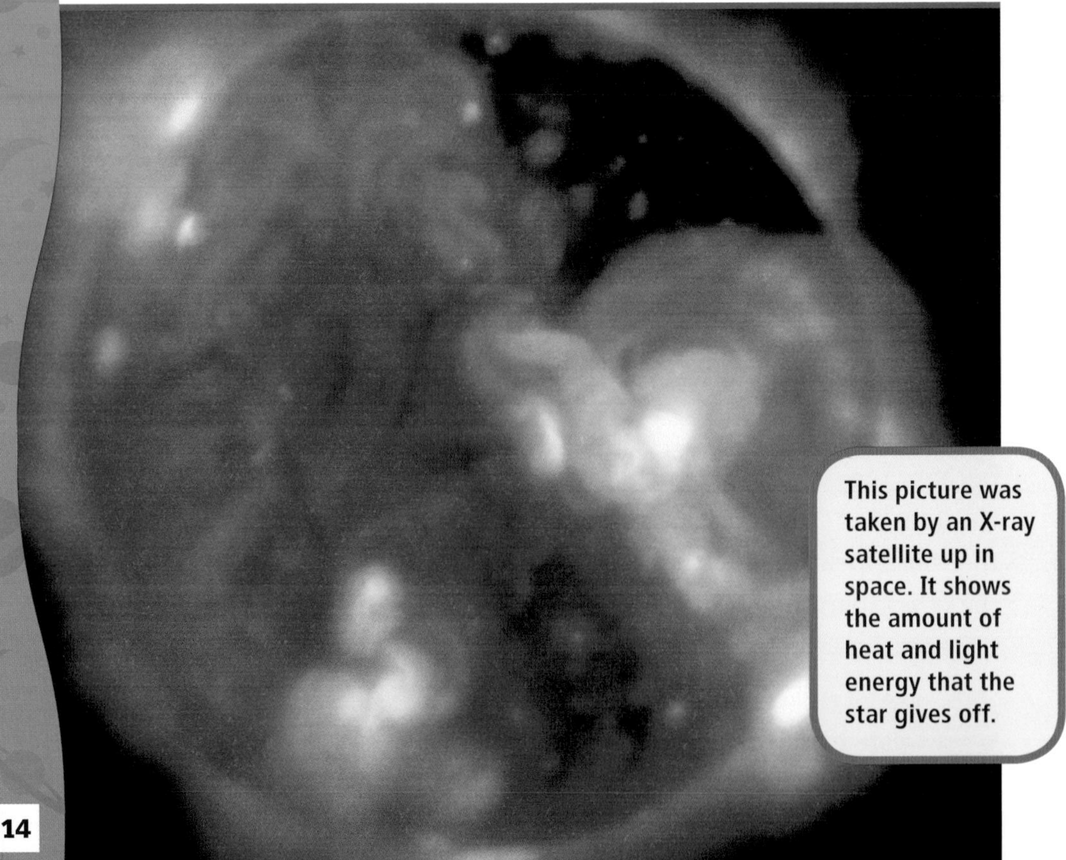

This picture was taken by an X-ray satellite up in space. It shows the amount of heat and light energy that the star gives off.

Astronomers learn a lot about space through the Hubble Telescope as it orbits Earth. It can detect things that Earth-bound telescopes can't.

Summary

The next time you look up at the night sky, you will know there is a lot more than meets the eye. You are not only part of Earth, but you are also part of the Milky Way galaxy. The sky is filled with young and old stars, as well as stars of different colors. The planets in our solar system all revolve around the sun but they are all very different. The inner planets are rocky and most of the outer planets are made up of gases. You also know that there are 141 known moons within our solar system. Who knows what new discoveries scientists will make about the universe?

Fast Fact

Venus shines brightly in the sky just before sunrise and after sunset. It's sometimes called the morning star or evening star, but it is still a planet. It shines brightly because the clouds surrounding it reflect a lot of the sun's light.

Glossary

asteroid (AS•ter•oyd) A piece of rock and metal that orbits the sun (8, 9)

comet (KAHM•it) A ball of ice, rock, and frozen gases that orbits the sun (10, 11)

galaxy (GAL•uhk•see) A huge system of stars (2, 3, 4, 15)

meteor (MEET•ee•er) A piece of rock, smaller than an asteroid, that enters Earth's atmosphere and burns up (9)

satellite (SAT•uhl•yt) A body in space that orbits a larger body (5)

Think and Write

1. Make a diagram showing the planets in order as they revolve around the sun in the center. Try to make the planets the proper relationship in size. You may refer to the textbook or other charts as you work.
2. **MAIN IDEA AND DETAILS** What kind of star is our sun as far as its size, color, and amount of heat and light it gives off?
3. **SEQUENCE** How do comets become meteors?

4. **Narrative Writing** Imagine that you are a newspaper reporter describing the discovery of a new planet. Write a description of the planet using elements from the selection. Include its size, whether its surface is made of rocks or gases, and whether it can sustain life. Be sure to include a name for this planet.

Hands-On Activity

You can see all of the colors in the sun's light with this activity. Place a tall glass of water in a sunny window. Put it on top of a sheet of plain white paper. Then take a piece of cardboard and cut a long vertical slit in the cardboard. Place it vertically between the glass and the window. When sunlight shines through the slit in the cardboard and through the water, it will split into the colors of the rainbow on to the white paper.

School-Home Connection

Join a family member outside on a clear night and see how many stars you can observe in the sky. Take note whether or not you can see the planet Venus in the sky.

GRADE 6
OL Book 11
WORD COUNT
1700
GENRE
Expository Nonfiction
LEVEL
See TG or go Online

Go online Harcourt Leveled Readers Online Database
www.eharcourtschool.com

ISBN 978-0-15-362271-7
ISBN 0-15-362271-7

Earth's Changing Ecosystems

Harcourt

Photo Credits: Robert Cameron/Getty; p. 2: John A Rizzo/Getty Images; p. 3: Harcourt; p. 4: Royalty-Free/Corbis; p. 5: Hank Morgan/Photo Researchers, Inc.; p. 6: David R. Frazier/Photo Researchers, Inc.; p. 7: Comstock Images/Getty Images; p. 8: Mc Donald Wildlife Photography/Animals Animals/Earth Scenes; p. 9: Jeremy Woodhouse/Getty Images; p. 10: Calvin Larsen/Photo Researchers, Inc.; p. 11: Rowan Beste/Animals Animals/Earth Scenes; p. 12: Royalty-Free/Corbis; p. 13: Peter Weimann/ Animals Animals/Earth Scenes; p. 14: Michael Thompson/Animals Animals/Earth Scenes; p. 15: Masterfile